# Planet & People

## LEAVING CERTIFICATE GEOGRAPHY

### Geoecology Option 7

Sue Honan

Sue Mulholland

MENTOR
BOOKS

**MENTOR BOOKS**
43 Furze Road
Sandyford Industrial Estate
Dublin 18
Tel: 01-2952112
Fax: 01-2952114
Website: www.mentorbooks.ie
Email: admin@mentorbooks.ie

| | |
|---|---|
| Edited by: | Treasa O'Mahony |
| Subject Editor: | Dr Tom Hunt |
| Book Design & Typesetting: | Nicola Sedgwick Kathryn O'Sullivan |
| Illustrations: | Michael Phillips |

ISBN: 978-1-84210-386-9

1 3 5 7 9 10 8 6 4 2

Printed in Ireland by Colourbooks Ltd.

# Contents

## GEOECOLOGY

# Acknowledgements

Brazilian Embassy; Clonakilty Agricultural College; Department of Agriculture; Jean Duffy; Tony Dunne; John Englishby; John and Maureen Enright; examsupport.ie; Dermot McCarthy; Con McGinley; Billy and Mimi McNabb; Anne Mulligan; *NewScientist*; Michael Redmond; Val Redmond; Scientific American; Staff and Students at St Laurence College, Loughlinstown; Staff and students of St Mary's College, Dundalk; State Examinations Commission; Jimmy Weldon; Anite White.

# Dedication

*For Chris, Eleanor and Maedhbh Honan*

*For Gerard, Patricia and Eoghan Mulholland*

# CHAPTER 1

# Soils

## At the end of this chapter you should be able to:

- **Name and describe the components of any soil.**
- **Explain how soil is formed.**
- **Name and describe the main characteristics of any soil.**

## Contents

## KEY THEME

**Soils develop from the weathering of rocks in situ and from re-deposited weathered material.**

# 1.1 Soil composition

What is soil made of?

Soil is composed of a number of ingredients/constituents.

1. **Mineral matter** – rock particles from the bedrock and weathered rock.
2. **Air** – found in the pore spaces between the rock grains.
3. **Water** – also found in the pore spaces between the rock grains. In dry weather, water forms a thin film around the grains. In wet weather, it fills the pores.
4. **Humus** – produced from decaying organic matter such as leaves and dead animals.
5. **Living organisms** – earthworms, beetles, fungi, bacteria.

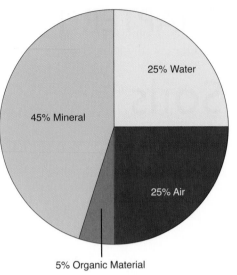

Fig. 1 Composition of soil

The components of soil are mixed in different quantities to create different soil types. Vegetation and time also play a role in creating soil. However, climate is the single most important factor in determining what a soil will be like as climate influences vegetation and the rate of weathering in an area.

## Soil formation

Several factors combine to create soil:

| | |
|---|---|
| 1. | Solid rock particles are broken down by mechanical weathering to form small soil grains. These make up the 'skeleton' of the soil. |
| 2. | Chemical weathering releases important nutrients from the rock grains, e.g. phosphorous, potassium, calcium. |
| 3. | Seeds are blown or carried onto the soil grains and may grow into plants that enrich the soil when they die. Early plants that can grow in young soils include mosses and lichens. |
| 4. | Micro-organisms decompose the remains of plants, which further enriches the soil. These micro-organisms improve the fertility of the soil, enabling a greater variety of plants to grow. |
| 5. | This cycle continues until the soil reaches its maximum fertility given the climate it is in. |

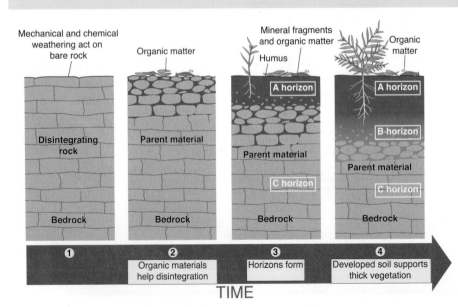

Fig. 2 The process of soil formation

2

# 1.2 Soil characteristics

When describing a soil, several factors are taken into account:

1. Colour
2. Texture
3. Structure
4. Humus content
5. The acidity/alkalinity of the soil (soil pH)
6. Water content/retention

## 1. Colour

Soils have a range of colours. Peat is dark black, other soils may be bright red or yellow. The colour of a soil depends on its parent material or on other processes that have occurred such as leaching.

## 2. Texture

The texture of a soil is controlled by the amount of sand, silt and clay particles in it. Different soils have different amounts of pore space, which affects soil aeration and drainage.

There are three main textures:

(a) clay
(b) sandy
(c) silty

When describing soil textures, the terms **silt**, **sand** and **clay** are important because they indicate the particular size of the mineral grains in a soil.

*Clay is the smallest. Its not visible to the naked eye.*

*Silt is next. It is barely visible to the naked eye.*

*Sand is the biggest, which is visible to the naked eye.*

| Type of soil | Description |
|---|---|
| Clay | Particle diameter is less than 0.002 mm. Not visible to the naked eye. |
| Silt | Particle diameter is between 0.002 mm and 0.05 mm. Barely visible to the naked eye. |
| Sand | Particle diameter is between 0.05 mm and 2 mm. Visible to the naked eye. |

Fig. 3 Description of soil grains

## Clay soils

Clay soil contains the smallest sized particles. Clay particles form a sticky soil when wet and will generally hold a shape after drying. Clay soils contain approximately **0-45% sand, 0-40% silt and 40-100%** clay. Clay soil is naturally high in nutrients so plants grow well. In summer, it is often baked dry with visible surface cracks, making it difficult to get water to the roots of plants. In winter, it can be constantly wet and waterlogging is common. At most times of the year, it is difficult to dig. If a soil feels sticky and smooth and it holds together like playdough, it is probably a clay soil.

## Silty soils

Silty soil contains particles which are smaller than sand particles but larger than clay particles. Silt feels powdery when rubbed between your thumb and forefinger. Silty soil sticks together when wet, but it will not hold its shape after it dries. Silty soils can be badly drained but do not often become waterlogged.

## Sandy soils

Sandy soil contains particles that can be seen with the naked eye and feels gritty when rubbed between your thumb and forefinger. Sandy soils tend not to stick together when wet. Generally they contain 85-100% sand, 0-15% silt and 0-10% clay.

Waterlogging is rare in sandy soils as they are very free-draining. However, watering and feeding of plants is needed on a regular basis because the nutrients drain away easily. Sandy soil is quick to warm up in the spring, so sowing and planting can be done earlier in the year than in clay or silty soil. If a soil feels gritty and sandy and does not hold together easily, it is probably a sandy soil.

## Loam soils

Soils are rarely composed of just sand, silt or clay. They are usually a mixture of the three with a larger percentage of one size of particles. Loam soils contain roughly equal amounts of clay, sand and silt. Most plants will grow in loam soils. It is brown and crumbly in texture and similar to that found in well-maintained gardens. This soil is rarely waterlogged in winter or dry in summer and supports a wide range of plants. Loamy soil is light and easy to dig and is naturally high in nutrients.

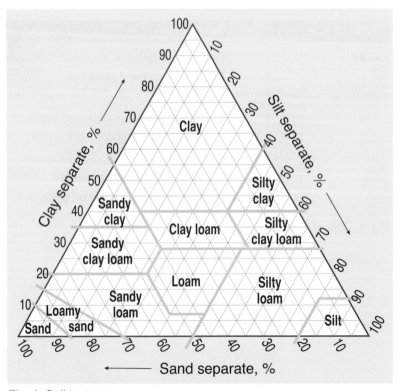

Fig. 4 Soil textures

| Appearance of soil under lens | Feel of soil between fingers | | Rolling of soil between fingers | Textures of soil |
|---|---|---|---|---|
| | Dry | Wet | | |
| Large (sandy) grains absent or very few in amount | Smooth and non-grainy<br><br>Slightly gritty | Generally very sticky<br><br>Plastic/pliable | Gives long threads which bend into rings, similar to the effect of rolling a strand of playdough between fingers | Clay<br><br>Sandy clay |
| Many sand grains present | Slightly gritty | Moderately plastic /pliable | Gives threads with difficulty which will NOT easily bend into rings | Clay loam<br><br>Silty clay loam<br><br>Sandy clay loam |
| Sand grains present but silt and clay predominating | Smooth | Smooth | Forms threads with broken appearance | Silty loam |
| Comparable proportions of sand, silt and clay | Gritty | Slightly plastic/pliable | Gives threads with great difficulty | Loam |
| Sand grains predominate | Gritty | Not plastic or pliable – only slight cohesion | Gives threads with very great difficulty | Sandy loam |
| Mostly sand | Very gritty | Forms a flowing mass | Does not give threads | Loamy sand<br><br>Sand |

Fig. 5  Summary table of soil textures

## 3. Structure

Soil structure is a description of the way in which soil grains are lumped together by humus and clay particles. If you pull a plant from the ground, its roots will hold soil grains. Look closely and you will see that the grains are in small lumps which are called **peds**. The shape of these peds indicates the structure of the soil.

The spaces between the peds hold air and water and are important for plants to access air and water in the soil. Overcropping and overgrazing damage the structure of the soil, reducing its ability to support plant growth.

**Some common soil structures**

1. Crumb/granular
2. Blocky
3. Platy

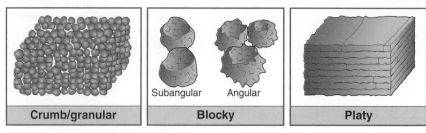

Fig. 6 Three common ped shapes

## 4. Humus content

The humus content of soil is important due to its ability to support plant growth. Soils rich in humus tend to be dark, with a good crumb structure. Humus is a dark black gel formed from rotting plant and vegetable material in a soil. This gel helps to bind soil grains together. As it is washed into the ground by rain, it adds nutrients to the soil. Living things in the soil also add humus to the soil with their droppings. In addition, earthworms, beetles and insect larvae burrow through the soil and create air spaces for plant roots. When these creatures die, their bodies decompose and add nutrients to the soil. Micro-organisms such as fungi and bacteria add nutrients by decomposing dead material.

## 5. Acidity/alkalinity of the soil (soil pH)

The pH of a soil is a measure of how acidic or alkaline it is. The acidity/alkalinity levels control which plants and animals will live in the soil. Most plants prefer a slightly acidic soil with a pH of 6.5. Peat soils are extremely acidic and contain few living things within them. Neutral soils are most suitable for bacteria which help to release nutrients such as nitrogen into the soil.

A fertile agricultural soil has a pH of 6.5 and may be composed of:
45% mineral particles (equal amounts of clay, sand and silt grains)
5% humus
25% water
25% air

## 6. Water content/retention

The amount of water a soil can hold depends on the humus content, texture and structure. Soils rich in humus can hold more moisture than those which lack humus.

Soils that have a sandy texture are often dry because water drains between the sand grains very quickly. A clay soil tends to hold more water due to the very small grains which have a large surface area and therefore hold more water around them.

Soils with a platy structure tend to become waterlogged as water cannot easily pass through this type of structure and gathers between the layers of the platy peds.

# questions

## Chapter Revision Questions

1. List and describe the main constituents (ingredients) of soil.
2. Explain how soil forms. In your answer refer to:
   (a) Mechanical and chemical weathering
   (b) Plant growth
   (c) Micro-organisms
3. Draw a table to summarise the characteristics of clay, silty and sandy soils.
4. (a) Explain the term soil structure.
   (b) Name three soil structures you have studied.
   (c) Draw a diagram to show each soil structure named in (b) above.
5. How does humus content affect soil? Refer to water content and fertility in your answer.
6. Using the chart, classify soils with the following compositions:
   (a) 50% clay, 40% sand and 10% silt
   (b) 30% sand, 60% silt and 10% clay
   (c) 15% clay, 45% sand and 40% silt
   (d) Identify the composition of the silty loam soil at the point labelled X.

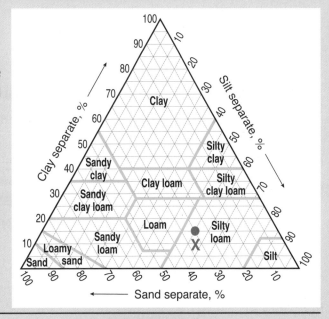

## Exam-style Questions

*soil componinti*

7. Examine the formation and texture of soils. [80 marks]
8. 'Soils can show a variety of texture and structure'. Discuss this statement. [80 marks]
9. 'Soils have a variety of characteristics.' Discuss this statement referring to four of the following:
   (a) Colour
   (b) Texture
   (c) Structure
   (d) Humus content
   (e) Acidity/alkalinity
   (f) Water content/retention [80 marks]

# CHAPTER 2

# Soil Forming Processes

**At the end of this chapter you should be able to:**

- **Name and explain eight processes which affect soil.**
- **Explain the influence that each of these processes has on the characteristics of a soil.**
- **Describe the range of soils across the world.**
- **Recognise different soils using diagrams of their profiles.**
- **Describe in detail the brown earth soils of Ireland and the latosols of Brazil.**

## Contents

## KEY THEME

**Soils are affected by their immediate environment and by a combination of processes operating in that environment.**

# 2.1 Processes that influence the characteristics of a soil

No two soils are alike in appearance or texture. In order to explain these differences, we must examine the processes that occur within a soil and the environmental factors that control these processes.

The major processes affecting soils are:

1. Humification
2. Leaching
3. Podzolisation
4. Gleying
5. Laterisation
6. Salinisation
7. Calcification
8. Weathering

**Climate, rock type, soil life and time all influence which of these processes dominates in a region**

The overall effect of the above processes is to form layers in a soil called **horizons**. (See page 14.)

## 1. Humification

This process releases nutrients into the soil. Humification is the method by which dead organic matter is converted into humus. Humus is a black gel composed of dead animals and plants. All dead plants and animals eventually decompose and become humus. Rain washes the humus into the soil where it is used by plant roots as a nutrient. Humification is important for soil as it makes it fertile.

## 2. Leaching

Rainwater washes soluble substances down through the soil. In very wet conditions, nutrients are washed from the soil altogether. In drier regions, minerals build up in a layer lower down in the soil. A certain amount of leaching is needed to wash humus into soil. However, excessive leaching is bad for soil because it makes it infertile.

## 3. Podzolisation

This is a type of leaching that occurs where rainwater is more acidic. This occurs due to the absorption of humic acids from rotting vegetation or by acidification due to pollution, e.g. acid rain. These soils form under coniferous forests and bogs. As these forests and bogs decompose, they add to the acidity of the rainwater.

The acidic rainwater dissolves all soil minerals except silica in the form of quartz. Quartz is a very resistant mineral. The top layer of podzolised soils is very pale due to the presence of quartz crystals. The layer below is enriched with the dissolved minerals and is darker in appearance. Podzols may also contain a layer of reddish iron oxide (rust). This is called a hard pan and can prevent water from draining through the soil.

## 4. Gleying

A gley soil is waterlogged for all or part of the year. The water table has risen into the soil and as a result, gley soils lack oxygen. Gley soils have patches of blue/grey colouration. They lack organic matter as little can grow in such wet, oxygen-poor (anaerobic) conditions. Gley soils are found in poorly-drained hollows. They are common in Ireland's drumlin belt (County Cavan, County Monaghan) and where the bedrock is impermeable (shale regions of County Clare).

## 5. Laterisation

This is a type of severe chemical weathering. It occurs in tropical and equatorial regions of the world. Leaching and high temperatures combine to dissolve all minerals except iron and aluminium oxides. These two minerals give the soils a red appearance and are known as latosols in Brazil.

## 6. Salinisation

Hot dry areas experience the reverse process to leaching. Evaporation causes salts in ground water to rise through the soil and build up in the upper layers. Salt is deposited on the surface as a hard white crust. If the salt concentration becomes too high, plants are poisoned and die. Salinisation is a problem for farmers who have to wash the crust away or break it up before their crops can grow. Irrigation salinisation is caused by excess water from irrigation which raises the water table, bringing salt to the surface. Irrigation salinisation can be reduced by using less water on crops and by growing crops which require less irrigation. Irrigation salinisation in Australia is estimated to cost the farming community €307 million each year.

Fig. 1 Salt deposits on the surface of a desert landscape

## 7. Calcification

This is the process by which calcium carbonate is concentrated near the surface of the soil. In regions of low rainfall, such as in the interiors of continents, the amount of water drawn up through the soil by plants (**transpiration**) may be greater than the precipitation falling on the soil. As a result of the imbalance between the amount of water plants draw up through the soil and the amount of precipitation falling on the soil, calcium carbonate builds up in the upper layer (A Horizon) of the soil. Calcium carbonate is a very useful substance for plants and these soils often have lush grass growth. When the grass dies, calcium carbonate is returned to the soil.

## 8. Weathering

Weathering is responsible for providing the mineral part of the soil. Soil grains are released from rocks by mechanical weathering such as freeze–thaw action.

Chemical weathering processes such as oxidation can release nutrients such as phosphorous, calcium and iron from the mineral grains. In the *Planet and People* Core Book (Chapter 6, page 77) the process of hydrolysis, which is responsible for releasing clay particles from granite rocks, is explained.

## 2.2 Factors affecting soil formation

Several factors can combine to influence which of the processes discussed on pages 9–10 occur. They are:

1. Climate
2. Relief
3. Parent material
4. Living things
5. Time

### 1. Climate

The development of any soil is closely linked to the climate of a region. Climate influences which soil processes are dominant in an area, e.g. soils in wet climates can be leached, gleyed or podzolised. In drier areas the soils may be salinised or calcified. Humification and laterisation are dominant in hot humid areas.

Climate controls the speed and type of weathering that takes place. In hot wet climates, such as in the tropical and equatorial zones, chemical weathering is rapid. As a result, deep soils develop. At the same time these are wet climates and leaching occurs. In colder regions such as the Tundra, the surface may be frozen for most of the year. Biological activity is slowed down and humification is delayed. Thin infertile soils may occur in these regions.

Fig. 2 In regions with cold climates, such as the Tundra, thin infertile soils occur.

Fig. 3 In tropical climates, deep soils can develop.

### 2. Relief

Relief can influence the depth and drainage of a soil. In general, sloping land is well drained and soils are quite dry. However, mass movement such as soil creep can occur so soils that develop on slopes are usually quite thin.

Flat upland areas are usually cold and wet. These conditions cause waterlogging. Because temperatures are low, the biological activity of animals and micro-organisms is

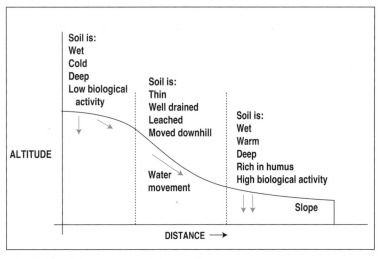

Fig. 4 The influence of relief on soils

slowed down so dead organic matter builds up but does not get converted into humus. Peat is an example of this. As a result, upland soils are often infertile.

Lowlands are warmer and usually flatter. This encourages the build-up of soil. Depending on the soil texture, they can be well drained. Biological activity is greater – earthworms and micro-organisms convert dead organic matter into humus. As a result, lowland soils are more fertile. However, local variations in aspect and relief may cause some areas to be colder or wetter, creating zones of waterlogged and boggy lowland soils.

## 3. Parent material

The type of rock that a soil develops from can influence its characteristics. The influences are summarised as follows:

(a) Igneous and metamorphic rocks tend to develop acidic soils.

(b) Sedimentary rocks can develop a variety of soils depending on rock type. For example, sandstone can produce acidic pale brown sandy soils, limestone produces calcium-rich dark-coloured soils, e.g. terra rossa, and shale tends to produce clay soils.

Fig. 5 Limestone and soil

The parent material of a soil is not always rock. In Ireland the majority of Irish soils have developed on glacial deposits of boulder clay, sands and gravels. These soils tend to be deep, fertile and well drained.

## 4. Living things

Living things within the soil can influence how the soil develops. As mentioned earlier, earthworms, beetles and insect larvae burrow through the soil and create air spaces for plant roots. Water can collect inside the burrows and keep the soil moist. When these creatures die, their bodies decompose and add nutrients to the soil. Micro-organisms such as fungi and bacteria add nutrients by decomposing dead material.

Plant roots help to anchor the soil to the ground and prevent it from being blown away. Leaves and branches protect the soil from the impact of 'rain-splash erosion' which in wet climates is very powerful and can remove tonnes of soil over a season.

## 5. Time

Time also plays a part in soil development. The longer that the soil forming processes listed above are in operation, the more developed the soil will become.

## 2.3 Soils of the world

Climate is the single most important factor in soil formation. This is because climate determines which soil forming processes will operate and how much biological activity can occur in the soil. Soils that have developed in response to particular climatic conditions are called **zonal soils**. Zonal soils are found in particular climatic regions, e.g. the soil associated with the cool temperate oceanic climate of Ireland is the brown earth soil.

| Climatic zone | Vegetation | Zonal soil |
|---|---|---|
| Tundra | Tundra | Arctic brown soil |
| Boreal | Coniferous forest (taiga) | Podzol |
| Cool temperate oceanic | Mixed deciduous forest | Brown earths |
| Mediterranean | Mediterranean | Red brown soils |
| Continental | Steppe/prairie | Chernozems |
| Tropical/ equatorial | Rainforest | Latosols |
| Desert | Desert | Aridisols |

Fig. 6 Zonal soils of the world

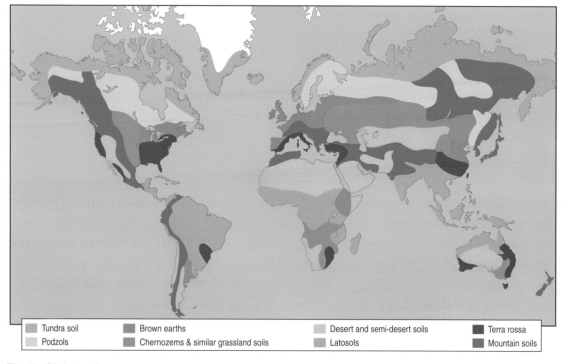

| | | | |
|---|---|---|---|
| Tundra soil | Brown earths | Desert and semi-desert soils | Terra rossa |
| Podzols | Chernozems & similar grassland soils | Latosols | Mountain soils |

Fig. 7 Global soils

Within the world's climatic zones there are variations in relief and local climates (micro climates). Due to these variations, factors such as unusual drainage, shelter and/or parent material may influence the zonal soil, changing it in some way. These altered soils are called **intrazonal** soils, e.g. gley soils.

Soils that are too young to have developed into full zonal soils are called **azonal soils**.

## 2.4 Soil profiles

A section of soil extending from the surface to the bedrock below is called a soil profile.

A soil profile may show different layers called **horizons** within the soil.

These layers have been created by a combination of processes, such as leaching./*lateralisation/weather + humification.*

A typical soil profile is shown below:

The **O Horizon** has a high percentage of organic matter, usually greater than 20% to 30%. Leaves, dead plants and animals collect here. This layer is rich in micro-organisms. In the lower part of this layer, humus is formed as the dead organic matter begins to decay.

The **A Horizon** is typically referred to as the topsoil. It is characterised by organic matter additions mixed with mineral materials. It is a dark horizon usually found at the surface or underneath the O horizon. An A horizon may have been altered to include properties of the O Horizon due to cultivation or other similar disturbances. This horizon has the most root activity and is usually the most productive layer of soil.

The **B Horizon** is also called the subsoil. This is a zone of accumulation where rainwater percolates through the soil and deposits material leached from above.

The **C Horizon** is not affected by the weathering process. The overlying soil horizons and materials often but not always develop from the C horizon.

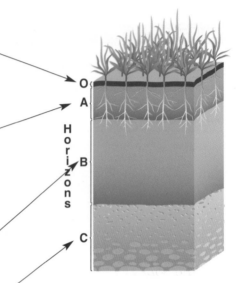

Fig. 8 Typical soil profile showing O, A, B and C horizons

Each soil has its own soil profile. These can be very detailed. In this chapter we will study two zonal soils: **Irish brown earths** and the **latosol soils of Brazil**.

### Irish soils

Ireland has a cool maritime climate and its natural vegetation is mixed deciduous forest. Therefore, its zonal soil is the brown earth soil. *because 3*

However, local conditions have modified this soil. Podzolisation has led to the formation of podzols, gleying has created gley soils and peat soils were created when wet conditions after the Ice Age prevented humification (rotting of plants into humus), thus allowing peat to build up. Therefore, a variety of soils are found across Ireland. See Fig. 9 on page 15.

*3l.    gleying has created gley soils along the Drumlin belt*
        *Cavan, Monaghan*
    *explain gleying.*

*explain humification here - tion...explain.
The process of humification...

## (a) Irish brown earth soils

Brown earth soil is a zonal soil. It has developed in response to our cool maritime climate and the presence of deciduous hardwood forest. In many areas, its parent material is boulder clay deposited during the last Ice Age.

This soil is very fertile due to the presence of humus (in this case, humus is composed of autumn leaves and dead animals) and a moderate amount of rainfall which washes nutrients into the soil but does not leach it.

As a result, this soil has no distinct horizons - it is uniformly brown in colour.

The pH of brown earth soil in Ireland varies from slightly alkaline to slightly acidic. This encourages the activity of bacteria, fungi and earthworms. These help to raise the fertility of the soil by drawing humus into the soil and decomposing it, releasing nutrients. The cool temperate climate is warm enough to allow biological activity to occur for more than nine months of the year. The brown earth soil has a good crumb texture that provides pore spaces for air and water, encouraging plant growth. As a result of its crumb structure, it is a well-drained soil.

Irish brown earth soils are highly productive and are used for tillage and pasture.

Soil distribution
% of total land area

29%
13.7%
0.7%
5.5%
5.1%
22%
24.0%

Mountain and hill blanket peat and podzolised soils
Mostly bare rock and rendzinas
Mainly blanket peat (low level and lithosols)
Basin peat
Mainly lowland gleys
Mainly acid brown earths and brown podzolics
Lowland gley brown podzolics and brown earths

Fig. 9 The variety of Irish soils

In Ireland, local conditions have created three variations in brown earth soil.

1.  Acidic brown earth soils occur on land 500 m above sea level on crystalline rock such as granite, schist or sandstone.

2.  Limestone and shallow brown earth soil (also called rendzina soil) occur in limestone areas such as the Burren in County Clare and in County Sligo.

Fig. 10 Brown earth soil in an Irish field

15

3. Podzolised (slightly leached) brown earth soils occur on the lowlands and on glacial drift. This type of soil covers 22% of the country.

*explain process of Podzolisation*

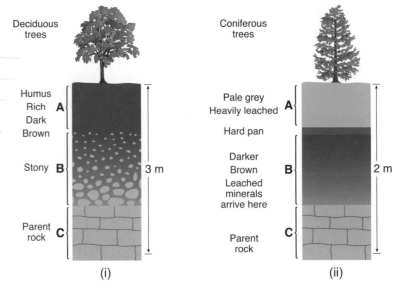

Fig. 11 Soil profiles of (i) brown earth and (ii) a podzol

## (b) A Brazilian soil type

### Latosols – Zonal soils of the equatorial region

Latosols are red- or yellow-coloured zonal soils associated with the equatorial climate. They occur in Brazil and other countries in the equatorial zone.

4. Latosols are the product of a hot wet climate where the process of laterisation is dominant. *showing the relationship between climate and soils*

These soils lack silica. They have a low humus content due to the rapid breakdown of organic material by bacteria which thrive in the hot and wet conditions of this region.

5. Leaching is so intense that only aluminium and iron compounds are left. These compounds give the soil its characteristic red or yellow colour.

6. Sometimes iron and aluminium compounds build up in a hard layer lower down the profile. If soil erosion removes the loose topsoil, the iron and aluminium rich lower layers are exposed. The high temperatures soon bake this soil into a hard brick-like surface which is impossible to cultivate even when wet. This type of soil is known as a **laterite**.

Latosols support the richest vegetation on the planet – the tropical rainforests, but the relationship between soil, climate and plant is a fragile one.

This relationship is possible because of the high rainfall and temperatures found in these regions. Fast plant growth combined with rapid decomposition of dead plants and animals forming humus allows the forest to use nutrients before they are leached out of the soil. So the 'nutrient cycle' is very short. Most nutrients are found in the thin upper layer of the soil. If the forest is removed, the soil becomes infertile within two or three years.

Fig. 12 Laterite soil – note its hard surface.

*Latosols → because of their hard brick-like surface are a major obstacle to the development of profitable agriculture in these regions.*

Latosols cover huge areas of Brazil, Africa and South East Asia. They are a major obstacle to the development of profitable agriculture in these regions.

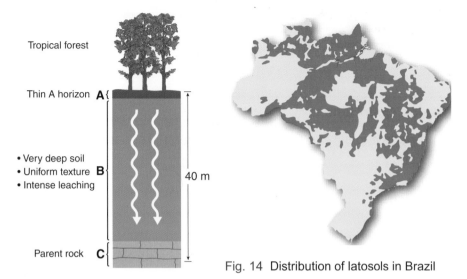

Fig. 13  Soil profile of a latosol

Fig. 14  Distribution of latosols in Brazil

Fig. 15  A boy breaking latosols into brick-shaped pieces.

The word laterite means brickstone. It is derived from the Latin word 'later' which means brick. For centuries, laterites have been used to build bricks in many different regions. For example, the temples at Angkor in Cambodia are built with laterite bricks and have survived 1,000 years.

# questions

---

**Chapter Revision Questions**

1. Name and briefly explain the following soil forming processes:
   (a) Humification
   (b) Leaching
   (c) Laterisation
   (d) Podzolisation
   (e) Calcification
   (f) Weathering
   (g) Salinisation
   (h) Gleying

2. Explain the impact of climate on soil formation.

3. Describe the impact of relief, parent material and time on soil formation.

4. (a) What are zonal soils? Give an example.
   (b) What is an azonal soil? Give an example.
   (c) What is an intrazonal soil?

5. Draw a labelled soil profile of a brown earth and a latosol soil.

6. Describe Irish brown earth soil. What processes have led to variations in this soil type?

7. Explain the factors that combine to create latosols.

8. Draw a summary table to show the similarities and differences between brown earth soils and latosols. Use the headings:
   (a) Colour
   (b) Depth
   (c) Soil processes
   (d) Climate/location
   (e) Fertility

---

**Exam-style Questions**

9. Examine the factors that influence soil characteristics.          [80 marks]
   *Leaving Cert Paper*

10. Compare and contrast the characteristics of any two soil types which you have studied.
    [80 marks]

11. Discuss the influence of parent material, climate and organic matter on soil formation.
    [80 marks]

# CHAPTER 3

# Soil Erosion and Conservation

## At the end of this chapter you should be able to:

- Name and explain the causes of soil erosion.
- Describe in detail how human activities can cause soil erosion.
- Name and explain the primary methods of soil conservation.

## Contents

## KEY THEME

Soil erosion is a challenge facing many countries. Human activities can cause soil erosion but several methods can be used to prevent it.

# 3.1 Human influence on soils

Humans have an important impact on soils. Human influence can trigger soil erosion due to poor farming methods and deforestation. Both farming and deforestation change soil characteristics and can damage the soil structure. In many regions of the world, human activities such as overgrazing and overcropping have led to desertification, soil erosion and famine. (See *Planet & People* Core Book, Chapter 11: Mass Movement Processes, Patterns and Landforms). Tourism in mountainous areas can also lead to the loss of soil as it is carried away on the boots of thousands of hillwalkers, e.g. in the Alps and Himalayas.

Soil is not a renewable resource. Once it is eroded, it is not renewed. Conservation of soil fertility and the prevention of soil erosion are challenges that face millions of people if their future food supplies are to remain secure.

## Soil erosion/degradation

Throughout the world, soil erosion is a serious problem. Each year the US loses about two billion tonnes of topsoil. In China, the loss of soil is a serious threat to that country's future food security. The loss of valuable topsoil due to wind and water erosion in the tropical lands of Africa is one of the greatest problems this continent faces.

Fig. 1 Soil is blown away from the land on this farm in the US.

Most soil erosion is caused by the wind and by water flowing downhill. It can happen on slopes with as little as a 10 cm drop in a 20 m length. The amount of erosion that occurs depends on:

1. The **quantity of water** moving downhill, e.g. more water = more erosion.

   The quantity of water moving downhill is affected by the amount of rainwater that falls and by the area over which it collects. The amount of rain cannot be influenced but the

Fig. 2 Intensive agriculture can lead to soil erosion.

   area over which it collects can be controlled by constructing barriers of some sort across the slope. (See soil conservation methods on pages 24–25 )

2. The **speed of the water**, e.g. faster flow = more erosion.

   The speed at which the water travels over the soil is affected by the slope of the land and by the length of that slope. The slope cannot be changed except by terracing the land, a process which takes a long time. However, the length of the slope can be controlled by constructing barriers at regular intervals.

3. The **state of the soil surface** and the type of soil, e.g. few plants, dry, compacted soil = more erosion.

   The third factor is the state of the soil surface and soil type. Soil type cannot be easily changed, although it can be improved by adding humus. However, the state of the soil surface can be influenced through cultivation and/or the amount of vegetation cover that is left or put on the land. Soils that are fertile and contain plenty of humus are less likely to be eroded.

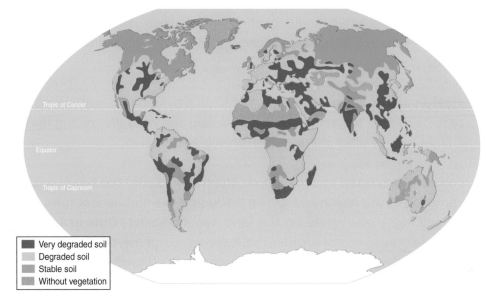

Fig. 3 Degraded soils areas of the world

## Desertification in the Sahel region of Africa

Desertification is the spread of desert conditions into new areas. It occurs in areas close to existing deserts. However, it can also happen in well-watered areas if the climate changes and becomes drier and human activities create conditions leading to soil erosion.

Desertification has two main causes:

1. **Human activities** such as overgrazing and deforestation, generally triggered by population growth.
2. **Climate change** – global warming is increasing drought conditions in certain areas of the world.

Fig. 4 This soil is too dry and compacted for food production to occur.

Fig. 5 The pink area on this map shows the Sahel region of Africa. Name the countries numbered 1 to 6.

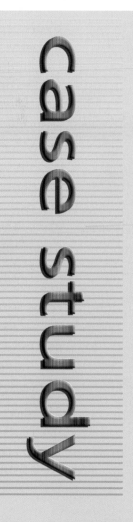

case study

## 1. Human Activities

In the Sahel region, soil erosion is occurring largely as a result of the human activities of **overgrazing, overcropping and deforestation.**

(a) **Overgrazing** occurs when farmers allow too many animals to graze an area of land. This damages soil structure and removes plant cover, allowing the soil to blow away.

(b) **Overcropping** occurs when land is continuously farmed. This drains nutrients from the soil and destroys the soil structure making it less fertile. The soil becomes dry, dusty and easy to erode.

(c) **Deforestation** occurs when large areas of forest are cut down, leaving a bare landscape. Forests provide natural protection for soil from rain and wind preventing mass movement. Tree and plant roots anchor the soil, preventing soil erosion.

(a) The human activities listed above damage the fragile soil of the Sahel region.

The population of the region has grown, including the numbers of farmers. As a result, the number of cattle and goats on the land has increased. **Overgrazing** of cattle has become a major cause of soil erosion in the Sahel. Large herds graze the land and their hooves compact the soil destroying its structure. Continued grazing also removes the protective cover of grasses increasing the risk of erosion.

(b) The growing population also needs more food from the already weakened soil. Despite improvements in farming methods, in some areas of the Sahel the land is being **overcropped**, i.e. the land is not left to rest between crops and/or is not fertilised properly. Because of population growth, the land around the edges of crop plantations is continuously in use. It cannot be left fallow to regain its nutrients as people would go hungry. The soil becomes overcropped, dry and dusty and blows away. Another reason why overcropping occurs in the Sahel region is the poor economic situation in many Sahelian countries. Many Sahelian nations availed of cheap loans from developed nations during the 1960s, but are now having trouble repaying these loans. They become Highly Indebted Poor Countries (HIPC). To qualify for debt relief, governments in the Sahel have to increase the amount of land under cash crops such as cashew nuts and cotton. These are grown on huge plantations and sold on the international market to pay off national debts. The plantation workers live on and use land on the

Fig. 6 Watering livestock at a rehabilitated borehole in southern Ethiopia

edges of the plantations to grow their own food crops. Therefore, growing cash crops allows desertification to encroach onto the plantations themselves.

(c) Deforestation means that the soil is easily eroded by the wind and infrequent flash floods. The increasing population is felling more trees for firewood and for building materials.

The lack of nutrients and destruction of soil structure renders the soil useless for future farming. Land is abandoned and over time the desert spreads onto these once productive farmlands.

The soil is also becoming drier due to the increasing demand for water for human and animal consumption and for irrigation. More wells are sunk and this, along with climatic changes, is causing the water table to drop.

Cattle dung, which was once left on the land to fertilise it, is now used as a fuel for cooking. This removes a source of valuable fertiliser, increasing the risk of soil erosion.

Due to the destruction of the soil, the region is now classed as **overpopulated** as it cannot feed its population.

## 2. Climate change

Global warming has changed the pattern of rainfall in the Sahel. It is causing a rise in the temperature of the atmosphere. As a result the air can hold more water vapour and condensation/precipitation is less likely to occur. The seasonal rains are becoming less reliable. Rainfall levels have decreased by as much as 30% over the last 20 years in the region. Droughts are therefore becoming more frequent and lasting longer. As a result, the soil is drying out and becoming exposed to erosion. Desertification is continuing.

Fig. 7 A desertified landscape in Africa

## Reasons why desertification is occuring

| | |
|---|---|
| 1. | In the Sahel region, population growth has led to an increase in cattle numbers on the land. This has reduced plant cover and compacted the soil. (Overgrazing) |
| 2. | Wells have been dug to provide water for cattle. Around the bore holes, so many cattle gather that the land is ruined over many hectares. |
| 3. | Plantation owners put huge areas of land under cash crops. This forces people to cultivate the land more often. Fields are not left to rest between crops. The soil loses nutrients and becomes useless. (Overcropping). |
| 4. | Rising populations create a demand for fuel wood. This further deprives the soil of important nutrients and increases soil erosion. |
| 5. | Drier climate due to climate change. |

Fig. 8 The above factors combine to cause soil erosion which leads to desertification in the Sahel.

# 3.2 Methods of soil conservation

## 1. Windbreaks

One important form of soil conservation is the use of windbreaks. Windbreaks are barriers formed by trees and other plants with many leaves. They are planted around the edges of fields. Windbreaks stop the wind from blowing soil away. They also keep the wind from destroying or damaging crops. They are very important for growing grains, such as wheat. Windbreaks can protect areas up to ten times the height of the tallest tree in the windbreak.

In parts of West Africa, studies have shown that grain harvests can be 20% higher on fields protected by windbreaks compared to those without such protection.

However, windbreaks seem to work best when they allow a little wind to pass through. If the wall of trees and plants stops wind completely, then violent gusts occur close to the ground. These gusts lift the soil into the air where it will be blown away. To avoid this, there should be at least two lines of plants in each windbreak. One line should be large trees. The second line can be shorter trees and other plants.

Fig. 9 Windbreaks prevent soil erosion in large fields.

Windbreaks not only protect land and crops from the wind, they can also provide wood products. These include wood for fuel and fences. Locally grown trees and plants are best for windbreaks as they are adapted to local conditions.

## 2. Contour ploughing/strip ploughing

With contour ploughing, the tractor operator follows the contours of a hillside, i.e. they go around the sides of the hills following the contours of the hillsides. This technique is more troublesome and potentially dangerous than the old method of going straight up and down the hillsides.

However, in terms of preventing soil erosion, the benefits of this technique are enormous. The furrows thrown up by the plough will now act as 'mini terraces', slowing or stopping the flow of rainwater and encouraging it to percolate into the soil.

By ploughing straight up and down the hillsides, the furrows act as ditches, enabling the water to flow down them picking up speed and soil. This fast-flowing mixture of water and soil increases the size of the furrow, deepening it until it reaches the harder subsoil below. Meanwhile, many tonnes of life-sustaining topsoil are carried away to the sea.

Contour ploughing is widely used in Europe and North America.

Fig. 10 A contour ploughed field

## 3. Stubble planting

This is suited to many of the grain and cereal crops planted in Ireland.

For centuries, the old stubble of crops harvested in a year was ploughed back into the soil. However, using the stubble planting method, the old stubble of harvested crops is not ploughed. Instead the stubble is left in place. Any fertilisers and new seed planted afterwards are inserted into the soil through small slits cut into the soil by a razor-type device attached to the tractor. In other words the soil is left virtually undisturbed.

The stubble left on the soil will rot into the soil eventually, increasing the humus content as it would have when ploughed in. However, the stubble will reduce wind and water erosion while the new crop is growing. It also gives cover and habitat to small birds. Stubble planting is used in the prairie lands of America and Canada. It is also used in Europe.

## 4. Terraces

Terraces work by reducing slope length and steepness to limit the energy of running water and its ability to carry soil away. Stopping or slowing the downhill flow of water allows the sediment to drop out of the water. This prevents gully erosion.

Terracing is used in Asia.

Fig. 11 Terracing on a hillside in Thailand

## 5. Stone walls or 'bunds'

These small walls are a very simple way of preventing soil erosion on a slope. These walls are placed along the contour of a hill and capture water, allowing it to filter into the soil rather than running off downhill.

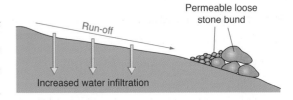

Fig. 12 Bund slows run-off allowing percolation of water into soil

## 6. Reduce ploughing in dry/windy weather

Not ploughing when the soil is dry and dusty reduces the risk of wind erosion. This is very important, because soil is lost during the ploughing process (see Fig. 13) and it is also lost if the following days are windy with no rain to dampen and settle the fine soil particles.

Fig. 13 Ploughing in windy weather increases soil erosion.

# questions

## Chapter Revision Questions

1. List three human activities that can lead to soil erosion.

2. Describe how the following factors influence the occurrence and speed of soil erosion:
   (a) Water
   (b) Soil type

3. Explain the terms:
   (a) Desertification
   (b) Overgrazing
   (c) Overcropping

4. Describe how human activities have led to soil erosion in the Sahel region of Africa.

5. How has climate change affected rainfall levels and patterns in the Sahel region of Africa?

6. Name and describe three methods used to prevent soil erosion. Draw labelled diagrams to illustrate your answer.

## Exam-style Questions

7. Examine how human activities can accelerate soil erosion.                [80 marks]
   *Leaving Cert Sample Paper*

8. Examine the causes of soil erosion and outline methods used to prevent it. [80 marks]

# CHAPTER 4

# Biomes

## At the end of this chapter you should be able to:

- **Explain the term** biome.
- **Describe in detail one major biome in terms of its climate, soil, plants and animals.**
- **Explain how animals and plants adapt to their biome.**

## Contents

## KEY THEME

The pattern of world climates has given rise to distinctive biomes. These biomes are world regions characterised by groups of plants and animals which have adapted to specific conditions of climate, soils and biotic interrelationships.

## 4.1 Biomes

Biomes are unique world regions which are controlled by climate. The climate of a region determines what type of soil is formed there, what plants grow there and which animals inhabit it. All four components (climate, soils, plants and animals) are interwoven to create the fabric of a biome. Examples of biomes include the tropical rainforest biome, the deciduous forest biome, the desert biome and the tundra biome.

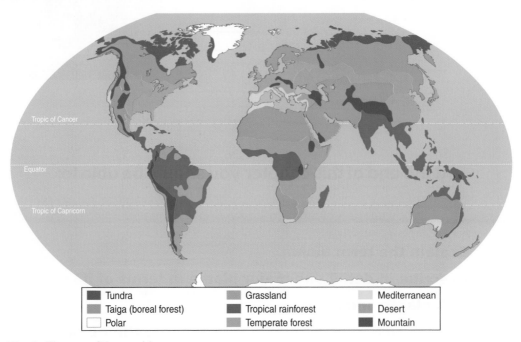

| Tundra | Grassland | Mediterranean |
| Taiga (boreal forest) | Tropical rainforest | Desert |
| Polar | Temperate forest | Mountain |

Fig. 1  Biomes of the world

## 4.2 The tropical rainforest biome

A tropical rainforest is a forest of tall trees found in a region of year-round warmth. An average of 1,250 mm to 6,600 mm of rain falls in a rainforest each year.

The tropical rainforest biome is home to the greatest variety of living things (biodiversity) on the planet. It is impossible to say exactly how many insects live in the rainforest, but one hectare may contain as many as 42,000 species.

Many fruits (bananas, citrus), vegetables (peppers, okra), nuts (cashews, peanuts), drinks (coffee, tea, cola), oils (palm, coconut), flavourings (cocoa, vanilla, sugar, spices) and other foods (beans, grains, fish) come from rainforests.

Fibres from tropical forests are used in rugs, mattresses, ropes, strings, fabrics, industrial processes and more. Tropical forest oils, gums and resins are found in insecticides, rubber products, fuel, paint, varnish and wood-finishing products, cosmetics, soaps, shampoos, perfumes, disinfectants and detergents.

It is estimated that a typical patch of rainforest measuring just 6 km² contains as many as 1,500 species of flowering plants, 750 species of tree, 400 species of bird, 150 species of butterfly, 100 species of reptile and 60 species of amphibian.

In short, rainforests are home to more species of plants and animals than all other regions of the world combined. The best-known area of this biome is the Amazon Basin of Brazil. The Amazon drains the world's largest tropical rainforest and is thought to contain about 3,000 species of fish, more than are contained in the entire North Atlantic Ocean.

## Location of the rainforest biome

The tropical rainforest biome is found between latitude 5° north and south of the equator.

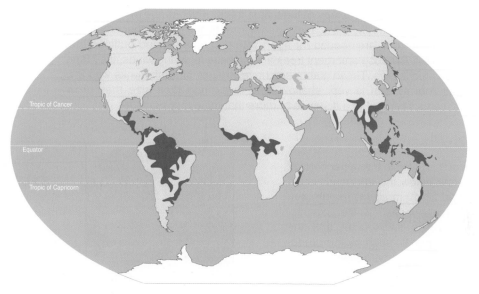

Fig. 2  The rainforest biome

The tropical rainforest can be found in three geographical areas around the world.
1.   Central America and the Amazon Basin.
2.   The Africa-Congo basin, with a small area in West Africa and in eastern Madagascar.
3.   Indo-Malaysia – off the west coast of India, Southeast Asia, New Guinea and Queensland, Australia.

## Climate

The climate in this region is tropical. The temperature in a rainforest rarely gets higher than 34 °C or drops below 20°C. The average temperature is about 27°C. There is a small temperature range. Average humidity is between 77% and 88%. Heavy convectional rainfall occurs each day in the afternoon. *an average of 1,250 mm to 6,600 mm of rain falls in a rainforest each year.*

As the sun moves from its position over the Tropic of Cancer (June 21st) to the Tropic of Capricorn (December 22nd), the area which receives the greatest amount of heat (known as the **Thermal Equator**) moves with it. This can cause a slight dry season at the edges of the forest zone.

# Vegetation

There is a great variety of rainforest vegetation. Each of the three largest rainforests (the South American, the African and the Asian) has a different plant group and animal species. However, the plant species look very similar and in many cases the trees can only be identified by their flowers.

There are four very distinct layers of trees in a tropical rainforest. These layers have been identified as the emergent, canopy, understorey and forest floor.

## 1. Emergent layer

The emergent layer consists of the tops of the tallest trees, which range in height from 40 m to 80 m. It contains birds like the scarlet macaw, insects and many other creatures. In the emergent layer trees are spaced far apart with umbrella-shaped outlines that grow above the forest. Because emergent trees are exposed to drying winds, they tend to have small, pointed leaves. These giant trees have straight, smooth trunks with few branches. Their root system is very shallow, and to support their size they grow buttresses that can spread out to a distance of more than nine metres.

## 2. Canopy

The canopy is the name given to the upper parts of the trees which grow below the emergent layer. The canopy is found 20-40 m above the ground. This leafy environment is full of life: insects, spiders, many birds like the toucan and the hornbill. There are mammals such as the orang-utan and the howler monkey (the second-loudest animal in the world after the blue whale).

The canopy is home to snakes, lizards and frogs. Plants in the canopy include thick, snake-like vines and **epiphytes** (air plants) like mosses, lichens and orchids which grow on trees.

## 3. Understorey

The understorey is a dark environment that is under the canopy. Most of the understorey of a rainforest has so little light that plant growth is limited. There are short, leafy, mostly non-flowering shrubs, small trees, ferns and vines (lianas) that have adapted to filtered light and poor soil. Animals in the understorey include insects (like beetles and bees), spiders, snakes, lizards, and small mammals that live on and in tree bark. Some birds live and nest within tree hollows. Some larger animals, like jaguars, spend a lot of time on branches in the understorey, looking for prey.

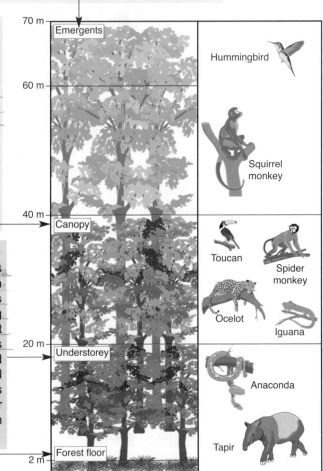

Fig. 3 Layers of trees in a tropical rainforest

## 4. Forest floor

The floor of the forest is teeming with animal life, especially insects and spiders like tarantulas. The largest animals in the rainforest generally live here, including gorillas, anteaters, wild boars, tapirs and jaguars. Indigenous people also live in the forest, e.g. the Yanomami of Brazil. Less than 1% of the light that strikes the top of the forest penetrates to the forest floor so the forest floor is often completely shaded, except where a tree has fallen and created an opening. The forest floor receives so little light that few plants can grow there. As a result, a person can easily walk through most parts of a tropical rainforest. The top soil is very thin and of poor quality. A lot of leaf litter falls to the ground where it is quickly broken down by termites, earthworms and fungi. The heat and humidity further help to break down the litter. This organic matter is then quickly absorbed by the trees' shallow roots.

Fig. 4  A scarlet macaw

Fig. 5  Jaguar in the rainforest

## Soils

1. The latosol is the zonal soil associated with the tropical rainforest biome. (See Chapter 2, pages 16 – 17.)

2. The latosol soil of the tropical rainforests is very poor in nutrients. Thousands of years of heavy rains have washed the nutrients away by the process of leaching.

As a result, the rainforest has a very short **nutrient cycle**. Nutrients are mainly found in the living plants and the layers of decomposing leaf litter on the surface (the O Horizon). Various species of decomposers such as insects, bacteria and fungi make quick work of converting dead plant and animal matter into nutrients. Plants absorb these nutrients the moment they are released.

A study in the Amazon rainforest found that 99% of nutrients are held in the root mats of the forest floor. When a rainforest is burned or cut down, the nutrients are removed from the ecosystem. The soil in the destroyed rainforest can only be used for a very short time before it becomes completely depleted of all its nutrients.

## Plant adaptations to the rainforest

Rainforest plants have made many adaptations to their environment. The adaptations can be grouped under the headings:

1.  Leaves
2.  Roots and trunks
3.  Parasitic plants

31

## 1. Leaves

With the high rainfall levels, plants have made adaptations that help them shed water off their leaves quickly so the branches don't get weighed down and break. Many plants have drip tips and grooved leaves. Some leaves have oily coatings to repel water. To absorb as much sunlight as possible in the dark shrub layer, leaves are very large.

## 2. Roots and trunks

Many emergent trees grow very fast in order to capture the light. They may have 'buttress' or 'stilt' roots for extra support in the shallow, wet soil of the rainforests (See Fig. 6). Trees have straight trunks that do not branch until a height of 30 m or more. This is because there is no need to grow branches below the canopy where there is little light. The majority of the trees have smooth, thin bark because there is no need to protect them from water loss and freezing temperatures. It also makes it difficult for plant parasites to get a hold on the trunks.

Many plants in the upper regions of the rainforest have aerial (air) roots. They are called **epiphytes**. The roots use the moisture in the air. Their sponge-like layers gather water and soak it up for later use. Plants like orchids, bromeliads and ferns have these types of root systems. They grow high in the canopy and the roots never reach the ground.

## 3. Parasitic plants

Parasitic plants live off the nutrients supplied by their host plant or use them for support. Vines or lianas are a good example. Lianas have adapted to the dark conditions on the forest floor by 'catching' a tree and taking a lift to the light. Lianas start off as small shrubs that grow on the forest floor. To reach the sunlight in the upper canopy, they send out tiny shoots to grab sapling trees. The liana and the young tree grow towards the canopy together. The vines can grow from one tree to another and may make up 40% of the canopy leaves. Strangler vines use trees as support and grow thicker and thicker as they reach the canopy, strangling their host tree which eventually dies. The network of vines look like trees whose centres have been hollowed out. Over 2,500 species of vines or lianas grow in the rainforest.

Fig. 6 A network of stilt roots support this rainforest tree.

## Animals of the rainforest

A tropical rainforest can contain more than 100 different species of animal in each hectare. The South American rainforest, particularly around the Amazon Basin, contains a wider variety of plant and animal life than any other biome in the world. The second-largest population of plant and animal life can be found in scattered locations around the west coast of India and on the islands of South East Asia.

There are many varieties of monkeys in the rainforests. In addition, different areas of the same rainforest may have different species.

Insects make up the largest single group of animals that live in tropical rainforests. They include brightly-coloured butterflies, mosquitoes, camouflaged stick insects and huge colonies of ants.

Fig. 7  There are many different species of monkeys living in each rainforest.

## Animal adaptations to the rainforest

Rainforest animals have made many adaptations to their environment. The adaptations can be grouped under the headings:
1.   Camouflage/colour
2.   Body structure
3.   Animal-plant associations

### 1. Camouflage/colour

Camouflage is one of the most effective adaptations used by a wide variety of animals. One of the most common and effective types of camouflage is looking like a leaf. The rainforest floor is scattered with dead leaves so animals and insects often use camouflage to look like dead or living leaves and are very hard to see from above. Moths and tree frogs use this method. Other animals such as the jaguar have spotted coats to blend into the shaded forest.

Animals also use colour to warn predators that they are poisonous. Some of the brightly-coloured animals are just bluffing. However, the poison arrow frog is most definitely not bluffing. This frog comes in many different colours - from sky blue to black and green. Poison arrow frogs get their name from rainforest tribes who use the secretions from their skin to poison the tips of their blow-gun darts.

Sloths are covered with a greenish layer of algae which camouflages their fur in their tree living (arboreal) environment. Sloths also move very slowly, making them even harder to spot.

## 2. Body structure

Other types of adaptation include the development of flaps of skin between the front and back legs. This allows some mammals such as 'flying foxes' to jump between the trees and glide for longer distances than they would normally be able to leap. Having a tail that can wrap around a tree branch (a prehensile tail) is another useful adaptation seen in animals such as lemurs.

Fig. 8 A tree frog in the rainforest

## 3. Animal–plant associations

There is a close relationship between plants and animals in the forest. Animals depend on plants for a home and food. In turn, plants depend on animals to fertilise and disperse seeds.

For example, the Ceiba tree is covered in vivid red flowers that attract many insects and hummingbirds, who drink the nectar, collect pollen and fertilise the tree. Some species of frog live only on one species of tree.

**ACTIVITY**

In your geography copybook, create a summary table of the ways in which plants and animals in the rainforest biome have adapted to their surroundings.

For more about the tropical rainforest biome, log onto
http://www.rain-tree.com/facts.htm

# questions

## Chapter Revision Questions

1. Explain the term biome. Name two biomes.

2. Draw an outline map of the world. Show the equator and the locations of the tropical rainforest biome.

3. Describe the climate of the tropical rainforest biome.

4. Name and describe the soil of the tropical rainforest biome.

5. Name the four layers of vegetation in the tropical rainforest biome.

6. Briefly describe the plants and animals found in each layer.

7. Describe how plants and animals have adapted to live in the tropical rainforest biome.

8. Describe the tropical rainforest biome under the headings:
   (a) Location
   (b) Climate
   (c) Vegetation
   (d) Adaptations of plants and animals to the environment.

## Exam-style Questions

9. Examine the characteristics of a biome that you have studied.
   *Leaving Cert Paper* [80 marks]

10. Illustrate the development of biomes, with reference to a specific example.
    *Leaving Cert Sample Paper* [80 marks]

11. Discuss the interaction between climate, vegetation and soil in a biome that you have studied. [80 marks]

12. Examine the adaptations of plants and animals to conditions in a biome that you have studied. [80 marks]

# CHAPTER 5

# The Impact of Human Activities on Biomes

## At the end of this chapter you should be able to:

- **Provide detailed examples of how people have changed biomes through their agricultural, industrial and settlement activities.**

## Contents

## KEY THEME

**Biomes have been changed by human activities.**

A variety of human activities have changed biomes across the world. These activities include:

1. Settlement
2. Deforestation and intensive agriculture
3. Industrial activity

In this chapter we will look at four human activities and their influence on various biomes across the world.

## Early settlement and the clearing of deciduous forest in Ireland

In prehistoric times a deciduous mixed woodland biome covered much of the landscape of Ireland. The most common trees were oak, elm, ash, Scots pine and alder.

Between 5,000 and 6,000 years ago, the first farmers began to clear woodlands to create farm land. Prior to this, there was only a small population of New Stone Age farmers and their stone tools took time to make and were not very strong. As a result, forest clearance was confined to small areas.

However, with the development of bronze and iron, forest clearance increased. Iron Age people had sharp and strong axes which increased both the area of land that could be cleared for settlement and farming and the speed at which this clearance could be carried out. From roughly 600BC to 500 AD, the amount of woodland in Ireland decreased. However, some trees and woodland plants were protected by law. The most important trees in Irish forests were the so-called 'nobles of the wood' – oak, hazel, holly, yew, ash, Scots pine and crab-apple. Under Brehon law, any person who damaged one of these trees had to pay a fine of two milk cows and one three year-old heifer.

case study

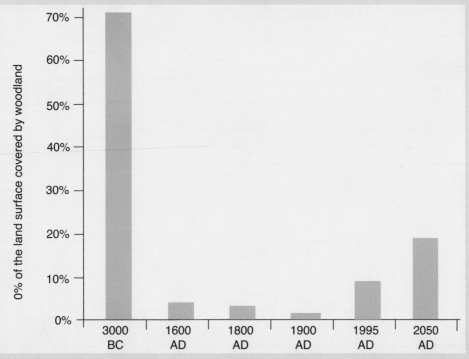

Fig. 1 Forest cover decline in Ireland from 1600 to 1900 and the projected rate of growth of forest cover up to 2050

Note: Time axis is not to scale

From 800 to 1000, the Vikings used wood to build settlements and transport (e.g. longboats) in Ireland.

From 1500 to 1700, the Norman settlement increased the amount of forest cleared. Timber was required by the Normans to construct bridges and defensive settlements.

In addition, English planters began to arrive in Ireland from the 1500s onwards, which further accelerated tree clearances due to the need for timber for building and industry.

Most of the remaining Irish oak forests were cleared at this time and used to build the English naval fleet. The population of Ireland was twice what it is today, so there was a greater amount of timber used for fuel.

By the early 1900s, only 0.5% (35,000 hectares) of the land area of Ireland was covered by woodland.

Today, roughly 12% of the land area of Ireland is under woodland: this is the lowest in Europe. The European Union average for woodland is 31%. It is hoped that we will have 16% of the land of Ireland under woodland by 2035. This will be a great achievement, but it will still leave us at just over half the EU average.

## Deforestation and intensive agriculture in Brazil

Human activities have had a massive impact on rainforests in the form of deforestation. We will examine their impact on the rainforest in Brazil.

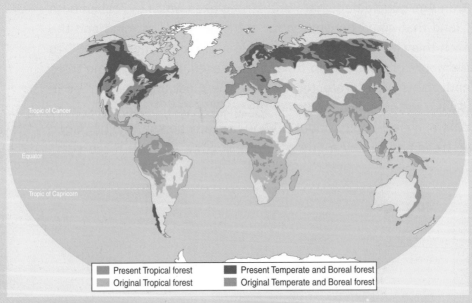

Fig. 2 Global distribution of original and remaining forests

## Causes of deforestation in Brazil

1. Intensive agriculture, e.g. soya plantations and cattle ranching.
2. Logging.
3. Demand for wood as fuel.
4. Construction of large dams and reservoirs drown forest.
5. Mining and industry clear forest to reach minerals.
6. Government–organised forest colonisation schemes clear forest for farming.

| ECONOMIC ACTIVITY | % DEFORESTATION % |
|---|---|
| Cattle ranches | 60 |
| Small-scale subsistence farming | 33 |
| Fires, mining, roads, dams, urbanisation | 3 |
| Logging | 3 |
| Other commercial agriculture | 1 |

Fig. 3 Economic activities leading to deforestation in Brazil

These activities are occurring because of the following underlying social and economic conditions in Brazil.

1. Overpopulation and poverty in Brazilian cities creates political difficulties.
2. The government wants to open up the Brazilian rainforest to take advantage of its timber and mineral wealth.
3. Beef producers require more land to herd massive numbers of cattle for the beef trade.
4. Brazilian industry requires more power if it is to develop and maintain its industrial strength.

Fig. 4 Logging in Brazil

The above four factors led to a situation in Brazil where the government began to encourage poor people from the cities to settle in the forest and clear it for farming. In some cases, land was given away. This actively promoted the destruction of the forest.

Timber companies were given rights to remove forest and sell timber abroad. The Brazilian government receives a portion of the timber companies' profits and uses it to pay off international debts.

Illegal growing and felling of timber is also leading to rapid deforestation, fuelled by a demand for cheap supplies of plywood and tropical timber locally and abroad. Illegal timber is estimated to account for 80% of all timber produced in the Brazilian Amazon.

As the area is rich in resources, licences were given to mining companies to clear forests and mine for metals such as iron ore and copper.

Roads are being built across the Amazon rainforest to allow access to logging companies, mineral exploration companies, soya plantations, cattle ranches and hydroelectric power (HEP) stations. To construct these roads, large tracts of forest were cleared.

The longest road is the Trans-Amazonian Highway, a 5,300 km road, built across Brazil from east to west. This highway was designed to facilitate settlement and exploration of resources in this vast under-populated river basin. It has allowed the movement of people and goods to previously inaccessible areas. Settlement soon followed upon completion of the highway's construction.

A new capital city, Brasilia, was built from scratch during the 1950s and 1960s in the heart of the Amazon rainforest. This was to encourage the settlement of the region. Today it has a population of 2.3 million. To construct this city, large areas of forest were cleared. More deforestation occurred on the outskirts of the city where small, temporary housing settlements were built for the migrant workers who moved to this area in order to construct the new capital city. Instead of returning to their original homes upon completion of the city, these workers chose to stay and avail of the greater opportunities in Brasilia.

An unlimited water supply and ideal river conditions led to the development of many HEP stations. Over 125 new HEP dams have been built in the Brazilian rainforest area. One example is the Tucuri Dam which caused over 2,500 km² of rainforest to be flooded. More than 8,000 people lost their homes and thousands of animals died.

In January 2000, the Brazilian government announced its plans for *Avança Brasil* (Advance Brazil). This is a €30 billion plan to cover much of the Amazon rainforest with 10,000 km of highways, hydroelectric dams, power lines, mines, gas and oilfields, canals, ports, logging concessions and other industrial developments.

Scientists predict that these planned developments will lead to the damage or loss of roughly 40% of Brazil's remaining Amazon rainforest. However, the government is finding it difficult to raise the money needed for these projects.

There are vast areas of the Amazon rainforest still intact. However, there are now 7,595 companies registered in the Brazilian Amazon and deforestation rates are growing at an alarming rate. This region now produces some 30 million cubic metres of logs a year or 90% of Brazil's total tropical timber production.

### DEFORESTATION FIGURES FOR BRAZIL

| Year | Deforestation (sq km) | Year | Deforestation (sq km) |
|------|------|------|------|
| 1990 | 13,810 | 1998 | 16,840 |
| 1991 | 11,130 | 1999 | 17,259 |
| 1992 | 13,786 | 2000 | 19,836 |
| 1993 | 15,410 | 2001 | 18,130 |
| 1994 | 14,896 | 2002 | 25,500 |
| 1995 | 29,059 | 2003 | 24,130 |
| 1996 | 18,160 | 2004 | 26,129 |
| 1997 | 13,040 | 2005 | 18,772 |
|  |  | 2006 | 13,100 |

Fig. 5 Rates of deforestation in Brazil

## The impact of intensive agriculture on the tropical rainforest biome

Intensive agriculture, especially the intensive production of soya beans, is also having a major impact on the tropical rainforest biome. Intensive agriculture completely disrupts the natural ecological balance of a biome.

The primary effects of intensive agriculture on the rainforest biome are:
1. Deforestation
2. Destruction of natural habitat
3. Introduction of exotic species

The expansion of intensive agriculture (soya plantations) in Brazil is contributing to the loss of 40 hectares of rainforest every minute day and night. This rate of forest clearance is driven by the increasing worldwide demand for soya and its products.

The destruction of the natural rainforest by intensive agriculture occurs in three ways:
1. Soils are damaged. Intensive agriculture has a huge impact on tropical soils because of the increased use of agrochemicals and mechanisation which can lead to soil compaction and also soil erosion.
2. Animals and plants cannot survive in the plantations and so the natural ecological balance that existed is disrupted or completely destroyed.
3. Plants grown in plantations are not native to the rainforest biome. These huge monoculture soya plantations bring a further threat to the natural habitat with the introduction of genetically modified organisms that have the potential to escape and invade natural communities.

Fig. 6 Deforestation for soya plantations in Brazil

All of the above contribute to a loss of biodiversity. Agricultural intensification has also led to the construction of waterways, roads and railroads which have contributed to the destruction of the biome.

**Impact of clearance on people and wildlife**
Rainforests are disappearing at about 40 hectares per minute, day and night. This clearance is having a significant effect on the biome and its people.

Fig. 7 Logs being towed down the Amazon River in Brazil

1. The area that was home to many native Amazonian Indians is greatly reduced. Their rights have been neglected. It has been suggested that some may have been murdered for trying to resist the clearance of the rainforest by ranchers and forest companies. Workers for the mining and forest companies spread diseases such as the common cold and measles. These diseases have killed thousands of native Indians as they have never been exposed to these germs before and therefore have no immunity to them.
2. Before 1500, there were approximately six million native people living in Amazonia. By

2000, there were less than 250,000. By the 21st century, 90 tribes of native peoples have been wiped out in Brazil alone.

3.  The area of natural habitat for wildlife is severely reduced. Many animals in the rainforest have not been clearly identified yet and as more of the rainforest is destroyed, the opportunity to study and identify these animals is lost.

4.  The loss of many species of plants is a serious cause for concern as some contain chemicals that could one day lead to cures for serious illnesses such as cancer and Aids. We already get many common drugs from different species of tree, e.g. **aspirin**. About one quarter of all the medicines we use come from rainforest plants. **Curare** comes from a tropical vine and is used as an anaesthetic and to relax muscles during surgery. **Quinine**, from the cinchona tree, is used to treat malaria. More than 1,400 varieties of tropical plants are thought to contain potential cures for cancer. These are being lost, cut and burnt at an increasing rate.

Fig. 8 The clearance of the rainforests in Brazil is having a significant impact on the biome and its people.

5.  Global warming. The loss of vast amounts of trees in the tropical rainforests will contribute to global warming. This will happen in two ways: first, the burning of forest adds $CO_2$ to the atmosphere. Second, by removing the forest we are destroying an important 'carbon sink'. A carbon sink is a thing or place where carbon dioxide is taken from the air and stored for a period of time. Plants act as carbon sinks as they use $CO_2$ in the cells of their bodies.

**Results of the forest clearance on Brazilian soils**

When a forest is cleared, the nutrient cycle is destroyed. The remaining soil can be easily washed away by heavy rain. In addition, as a result of the high temperatures in this region, the exposed soil is baked into a hard, brick-like surface which cannot support plant growth. This is known as a **laterite** soil, which is useless for farming. Settlers who had been persuaded by the government to leave the cities and settle in these areas find that the land they had hoped to work is useless. Many move back to the cities as a result.

The grass growth on the latosols is so poor that the beef cattle do not thrive and even more land is cleared to feed them.

# Industrial activity and European forests

Industrial activity is a primary cause of **acid rain**.

The acid in acid rain comes from two types of air pollutants: sulfur dioxide ($SO_2$) and nitrogen oxides ($NO_x$). These are produced by power stations burning fossil fuels as well as by car, truck and bus exhausts.

When these pollutants reach the atmosphere they combine with water in clouds and change to sulfuric acid and nitric acid. Rain and snow wash these acids from the air onto the land.

Acid rain refers to all types of precipitation such as rain, snow, sleet, hail and fog that is acidic in nature. Acidic means that these forms of water have a pH lower than 5.6, which is the pH of average rainwater.

Acid rain kills or damages trees, aquatic life, crops, other vegetation, buildings and monuments. It corrodes copper and lead piping, reduces soil fertility and can cause toxic metals to leach into underground drinking water sources.

Acid rain is a particular concern for coniferous forests in Scandinavia and Eastern Europe. It is estimated that more than 65% of trees in the UK and over 50% of trees in Germany, the Netherlands and Switzerland are affected by die-back due to acid rain.

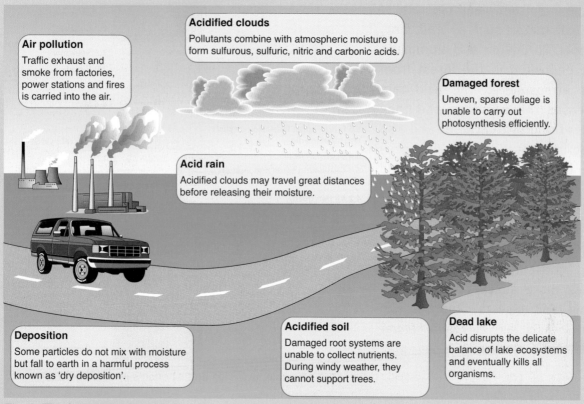

Fig. 9  The formation of acid rain

Fig. 11  Trees affected by acid rain

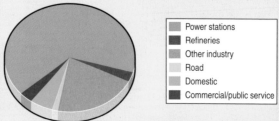

Fig. 10  Sources of sulfur dioxide emissions

## The effects of acid rain on forests

Some of the most important effects of acid rain on forests involve nutrient leaching, the accumulation of toxic metals and the release of toxic aluminium into the water and soil.

Nutrient leaching occurs when acid rain displaces calcium, magnesium and potassium from soil particles, depriving trees of these important plant minerals.

Toxic metals such as lead, zinc, copper, chromium and aluminium are deposited in the forest from the atmosphere. The acid rain releases these metals and they stunt the growth of trees and also that of mosses, algae, nitrogen-fixing bacteria and fungi needed for forest growth.

Trees are harmed by acid rain in a variety of ways. The waxy surface of leaves is broken down and nutrients are lost, making trees more susceptible to frost, fungi and insects. Root growth slows and as a result fewer nutrients are taken up.

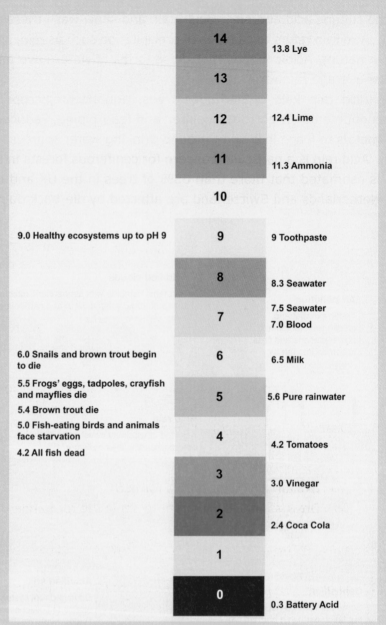

Fig. 12 A pH scale: acid rain and its effects on wildlife

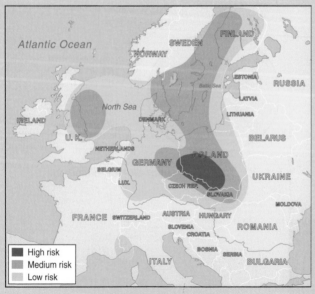

Fig. 13 Areas at risk from acid rain in Europe

# questions

---

**Chapter Revision Questions**

1.  Name three human activities that can change biomes.

2.  Briefly outline how the spread of historic settlement led to the clearance of the deciduous forest biome in Ireland.

3.  Name three causes of deforestation in Brazil

4.  Explain how these three causes resulted in deforestation in Brazil.

5.  What effect has intensive agriculture had on the tropical rainforest biome?

6.  Describe the impact of rainforest clearance on people, wildlife and soils.

7.  (i)   What is acid rain?
    (ii)  Explain how acid rain is formed
    (iii) Draw a labelled diagram to show the formation of acid rain.

8.  What is the effect of acid rain on the coniferous forest biome of Europe?

9.  Look at the table below: With reference to the table describe the trends shown.

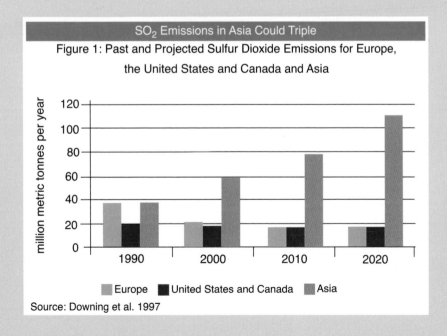

Figure 1: Past and Projected Sulfur Dioxide Emissions for Europe, the United States and Canada and Asia

Source: Downing et al. 1997

10. Look at the diagram showing the pathways that acid rain might take.
    Explain what is happening at positions A to G.

**Exam-style Questions**

11. Examine the impact of human activity on a biome that you have studied.   [80 marks]
    *Leaving Cert Paper*

12. Assess the global implications of the continued felling of tropical rainforests.[80 marks]
    *Leaving Cert Sample Paper*

13. 'A combination of factors is leading to the rapid loss of forest in Brazil.'
    Discuss this statement in detail.                                       [80 marks]

14. 'Economic development has an impact on biomes.'
    Discuss this statement with reference to a biome that you have studied.   [80 marks]

# Index